A THIEF IN THE MEADOW

G. Katz Chronicle

Publisher: IMNF Education Press, Pleasant Valley, NY

Library of Congress Control Number: 2023923368

A Step-up Companion Book to *A Thief in Pollinator Paradise*

Cover Design: G. Katz Chronicle

Thank you to Allyson Levy and Scott Serrano of Hortus Arboretum and Botanical Gardens for access to their beautiful greenery and their plant expertise.

Thank you to Francis Groeters of Catskill Native Nursery for his feedback and access to his extensive native plant collections.

As always, special thanks to Sabrina Kee for all of her technical and creative assistance.

Thank you to Bruce Fister for his editorial suggestions.

Prologue

You know that animals can move, react, and communicate—but do you know that plants can too? They just do it differently. The story you are about to read shows some meadow dwellers (“Meadowians”) and the conversations they might have.

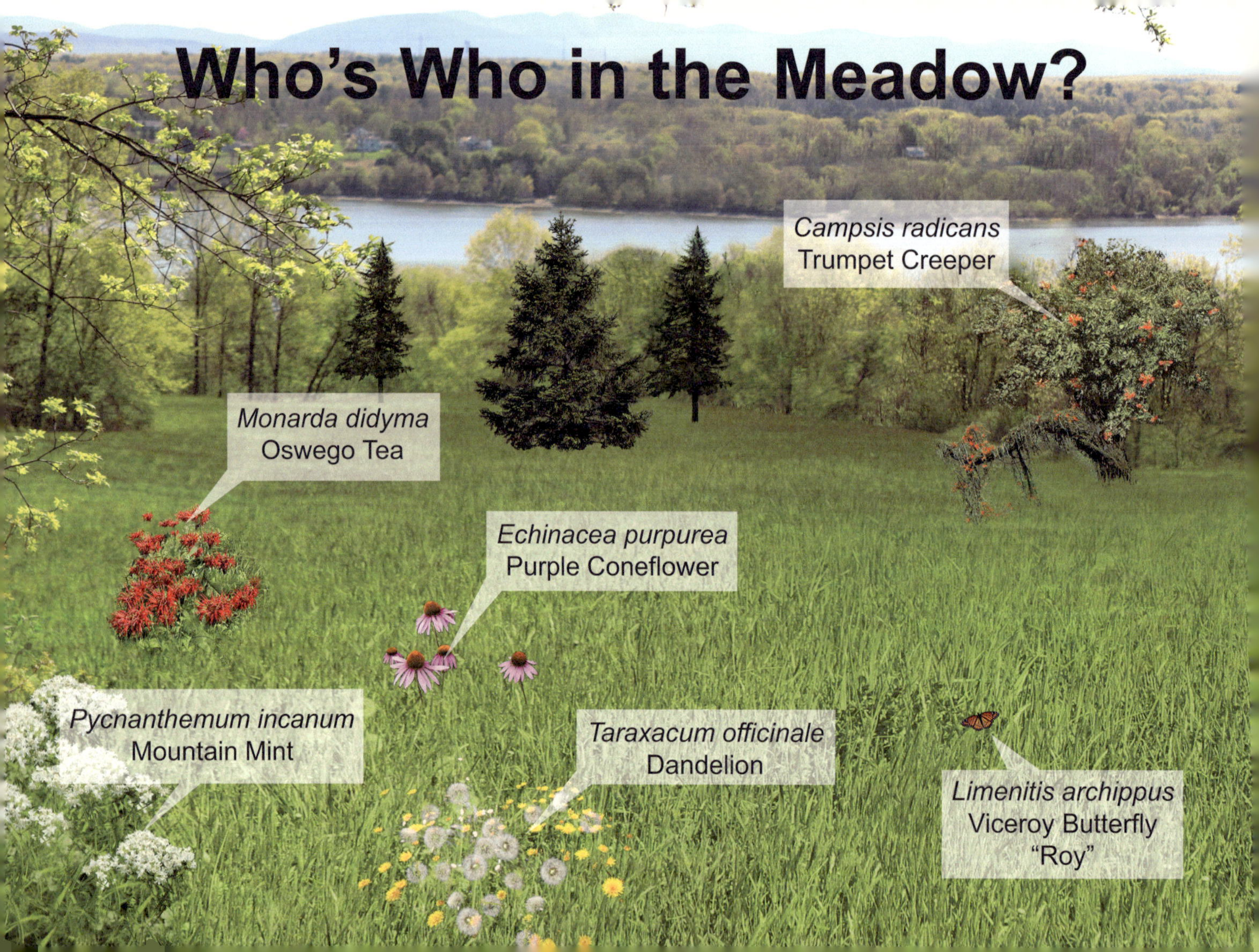
Who's Who in the Meadow?
Campsis radicans
Trumpet Creeper
Monarda didyma
Oswego Tea
Echinacea purpurea
Purple Coneflower
Pycnanthemum incanum
Mountain Mint
Taraxacum officinale
Dandelion
Limenitis archippus
Viceroy Butterfly
"Roy"

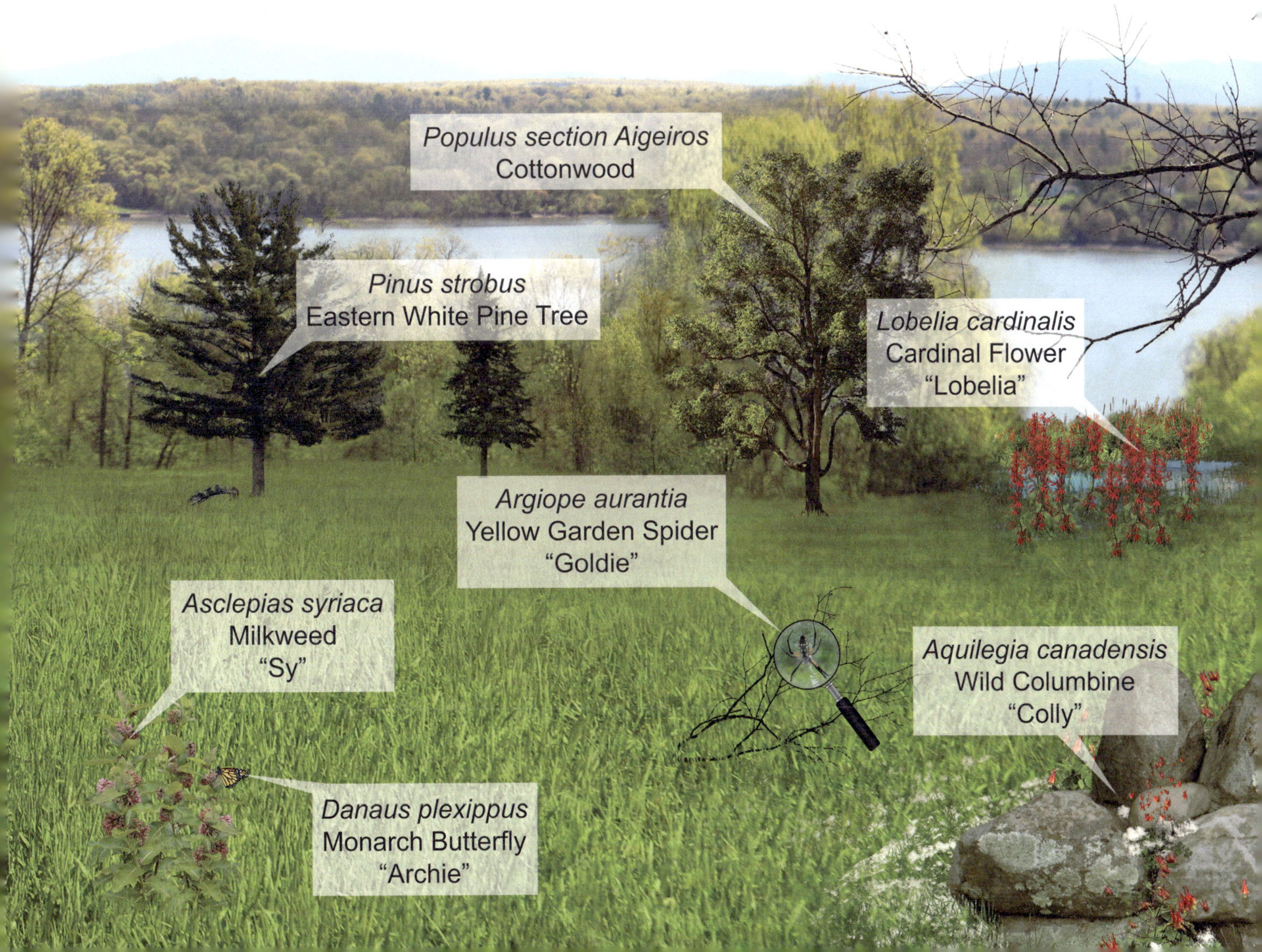

Populus section Aigeiros
Cottonwood
Pinus strobus
Eastern White Pine Tree
Lobelia cardinalis
Cardinal Flower
"Lobelia"
Argiope aurantia
Yellow Garden Spider
"Goldie"
Asclepias syriaca
Milkweed
"Sy"
Aquilegia canadensis
Wild Columbine
"Colly"
Danaus plexippus
Monarch Butterfly
"Archie"

BZZZZ

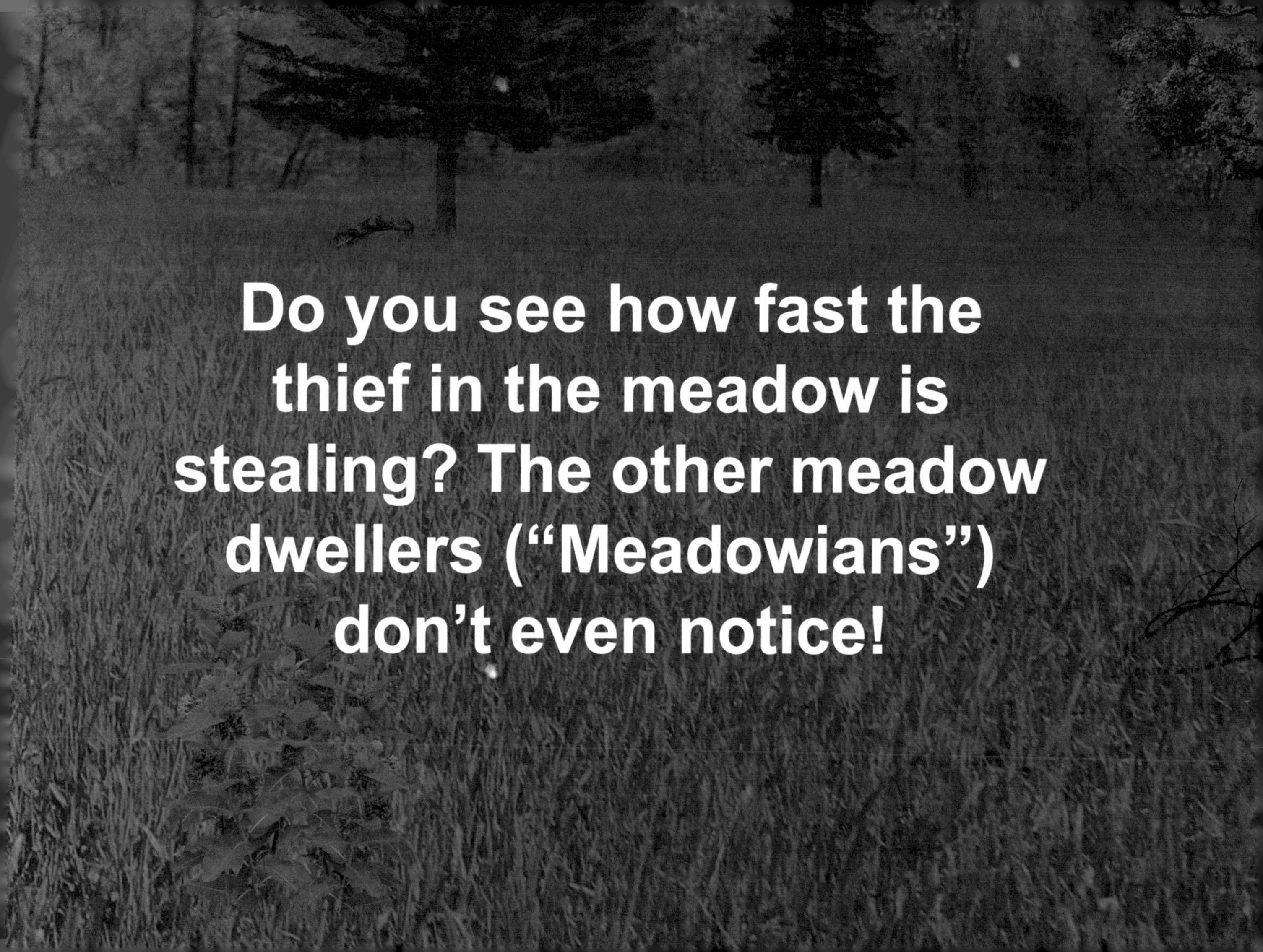
Do you see how fast the thief in the meadow is stealing? The other meadow dwellers ("Meadowians") don't even notice!

CRACK! The big pine dropped one of its branches in the night. The meadow flowers resting nearby sensed the disturbance.

“Did you hear that crack last night, Lobelia?” asked Colly.

"Yeah…I think that big pine over
there dropped one of his branches,"
said Lobelia.

Despite the noise of the preceding evening, the new day saw the usual rush of business in the meadow.

BZZZZ...
BZZZZ...
BZZZ

The next day, like all days, was busy. Flowers photosynthesized. Birds called out, claiming their territory. Bees and butterflies searched blossoms for nectar and cross pollinated the flowers along their journey.

Most flowers were happy for the visits.

But Lobelia was
not happy.

She did not welcome Roy, the Viceroy butterfly, as he approached.

BZZZZ...

But when Archie, the Monarch butterfly, floated over, she offered her nectar to him

After Archie sampled Lobelia's nectar and pollinated her flowers, he floated over to sip some of Sy's nectar.

Sy greeted him, “How are you old friend?” Asked Sy.

“I am well, thanks to you!” Replied Archie.

"You're one of just a few organisms that can feed on me." Sy laughed.

"Yes," agreed Archie, "The nourishment I get from you makes my predators sick, so they leave me alone!"

As Archie floated off, Sy and Lobelia cheered with gratitude, "My future offspring thank you!" Lobelia called after him.

Through all the photosynthesis, bird calls, nectar probing, and pollinating, there was a soft buzz in the meadow. The Meadowians would have heard it, if they stopped to listen. But they were too preoccupied with their daily business.

The next day, there was a flurry of activity. As Roy was taking his flap, flap, glide, flap, flap, glide around the pond, the frog in residence tried to eat him for breakfast. Fortunately for Roy, the frog tasted that Roy was toxic and let him go!

BZZZZ

Lobelia witnessed the attack, and with surprise, asked Roy, “Why did that bullfrog let you go? Were you too small of a meal for him?”

Roy started to speak but Goldie interrupted him, “Thief! Thief!” she exclaimed as she zipped about.

Switching his attention away from Lobelia, Roy asked, “What got stolen?”

“Part of my web! I was just about to rebuild the center when I noticed it drooping because someone took one of the anchors,” Goldie huffed as she crossed her two front appendages.

“Hmmm. Yes…I see your web is flying in the breeze. Maybe the wind blew some of it away,” suggested Lobelia.

"Nonsense," said Goldie. "I spin the strongest silk on this side of the Atlantic Ocean."

Helpfully, Roy assured her, "I'm sure we'll all be on the lookout for a web destroyer."

“Hmmpf,” replied Goldie as she hurried off to rebuild.

The sun continued shining, the breeze blew, the birds chirped, the bees buzzed. The meadow flowers grew, and there was a soft buzz in the meadow, if any of the Meadowians stopped to listen.

BZZZZ

BZZZZ

BZZZZ...

BZZZZZ...

"We had two unhappy Meadowians yesterday." Observed Sy, "Goldie's web was damaged and the bullfrog missed breakfast."

"Yes, why is that?" Demanded Lobelia, "Roy was on the frog's menu."

"Roy's orange markings are just as legit as Archie's." Explained Sy.

“Is that true?” Lobelia stared at Roy with astonishment, as she mentally compared his markings with those of Archie, the monarch butterfly.

"Oh, yes. As caterpillars we feed on toxic leaves. Surely that is what made the bullfrog spit me out!"

"I am *very* glad to know you haven't been deceitfully passing yourself off as a Monarch butterfly! I have newfound respect for you, Roy," Lobelia bobbed toward him gently in the breeze.

Roy unfurled his proboscis for the nectar which Lobelia was now offering. “On the contrary, Viceroy butterflies are survivors. We deceitfully pass ourselves off during each phase of our lives, sometimes even as waste. We live on, and reproduce. Like I’m helping you do now!”

As he deposited pollen from one flower to the next, Lobelia looked at her frenemy and said, “We have a truly symbiotic relationship, Roy.”

"Yes Lobelia. Roy the Viceroy and Archie the Monarch also help each other. Some predators leave them alone because their markings carry a warning of danger!"

BZZZZ...
BZZZZ...
BZZZZ...

BZZZZ...
BZZZZ...

BZZZZ
BZZZZ...

"Now I see that Roy and I have a symbiotic relationship, like all flowers and their pollinators," said Lobelia.

"I'm glad we have harmony," Said Roy, "But I've been noticing some of our symbiont neighbors are missing. There are fewer lichens than there used to be."

"That's odd," Responded Lobelia, "Fungi and algae don't have legs. They couldn't walk off!"

Said Sy, "Hmm, and they didn't finish their job, either. The fungi and algae grow as lichens on the rocks to break them down....yet the rocks are still there. That *is* a puzzle!"

"Come to think of it, earlier I noticed that there is less moss than there used to be," remembered Lobelia.

Goldie, who had been listening, scampered over and suggested, "Maybe the wind took the moss and lichens away."

"No way," Lobelia quickly answered. "There hasn't been anything more than a slight breeze during the last several days. Certainly not a strong enough wind to pull lichens or moss off of rocks."

"I agree," Goldie replied, "and I said the same when you suggested the wind took my web the first time it was destroyed. Since then, I've gotten robbed *again*."

"Twice?!" Came Sy's response. "I don't think the wind has been strong enough to disturb lichen or moss…and spider webs are tough enough to move in the breeze and not come apart." He thought for a moment and continued, "Let's consider this logically. Has our meadow environment changed in any way that might make it less habitable for lichen or moss?"

BZZZZ...
BZZZZ...

Lobelia, Goldie, and Sy all pondered the question a few moments.

After a bit, Lobelia, yawning, replied, “That is a question for another day, Perennials and Arachnid. It’s time to shut down photosynthesis for today and focus on respiration.” To Sy and Colly she said, “I suggest you two do the same.” To Goldie she said, “You do whatever arachnids do.”

"I hear that!" said Sy.

Goldie, always frustrated, shook her head and scurried back home.

BZZZ

Sy began the morning by saying, “What we know so far is that sections of Goldie’s web have gone missing twice, I can see that some lichens on these rocks are missing, and Lobelia can see that some moss near her is missing.”

Lobelia nodded in the breeze; Archie and Roy tilted their antennae and Goldie crossed her forelimbs and replied, “Hmph.”

Colly added, “I think that we need to do some investigation to find out if there is anything else missing from our meadow.”

"That sounds reasonable," agreed Lobelia.

"Yes, some research. Smart!" said Archie.

"Good thinking, Colly!" Roy chimed in.

Sy was delighted everyone was on the same page. He said, "I'm in!"

Just to be contrary, Goldie reacted, "How will you manage that, Sy? I can't remember the last time I saw you take a jog over to the big pine tree."

"Obviously, we need you guys to assist with our mission," Sy said to Roy and Archie, without reacting to Goldie's tone.

“Absolutely. Roy, glide across the field and then circle around to the river.” Archie continued, “I’ll float around the surrounding trees. We’ll meet back here with a report.”

With two flaps of his wings, Roy was on his way.

BZZZZ...
BZZZZ...
BZZZ...

BZZZZ

With a flap, flap, glide, flap, flap, glide, Roy arrived and immediately reported, “Lots of fluffy dandelion down and quite a few sections of cattails were noticeably absent in our meadow!”

"Oh!" reacted Lobelia.

"That's something," remarked Sy.

Archie floated over to the group and told everyone, "The big pine tree dropped a branch a few days ago and has been oozing resin to seal up the damage. Some of that resin has gone missing."

"Besides that," Archie continued, "all that fluff that carries the seeds of the cottonwood tree should have landed on the ground all over. Lots of it is missing."

“That’s a loud bee,” said Lobelia, changing the subject.

"It almost sounds mechanical. I saw a human once with a drone. Could it be that?" asked Roy.

"Oh! Look!" said Sy.

All now saw the Ruby-throated hummingbird headed for the Trumpet Creeper. She drank some nectar, pollinating the flowers as she moved among them.

And just as quickly as she appeared, she rushed back to incubate her eggs.

During the following days, the Meadowians paid better attention to Ruby's comings and goings as she gathered nectar and tended to her eggs.

DAY 1

DAY 7

Sy, Lobelia, Colly and all of their Meadowian friends found the tiny thief in their meadow, who was also the most nimble of pollinators: the Ruby-throated hummingbird.

DAY 8

DAY 14

AWW!

HOW
CUTE!

Did You Know?

A flower is a modified shoot, which serves as the reproductive organ of a plant; an inflorescence is a group of flowers arranged on a floral axis. *Echinacea purpurea*, *Asclepias syriaca*, and *Lobelia cardinalis* are actually inflorences, clusters of many tiny flowers.

Barn owls in New York also nest in tree cavities. They have a white, heart-shaped face, dark eyes, and long legs. They are found in meadow, grassland, or marsh areas where they feed on small mammals. They are nocturnal.

Birds steal from spiders. Many species of birds use spider silk when constructing their nests. The silk has the advantage of being strong while also being flexible. As the nestlings grow, the nest can stretch with them, and the silk allows for easy repair in the case of damage.

Campsis radicans is a favorite of hummingbirds. However, some people think its more trouble than its worth because it will vigorously grow over any kind of support and will live in nearly any conditions. An additional concern is hinted by its nick name, Cow Itch Vine; The Trumpet Creeper is a skin irritant.

Eastern Small-Footed bats (*Myotis leibii*) are known for their small feet and black facemasks. They forage at night in wooded areas, over streams and ponds, and along cliffs. Moths compose nearly half of their diet. They eat soft-bodied prey.

Evergreen trees grow scabs over their wounds. Both evergreen and deciduous trees use xylem and phloem cells for the transport of sap (sugar and water). Evergreen trees also have another liquid in their outer cells called resin. If an evergreen becomes injured, resin oozes out, covering the area and preventing infection. Resin is antimicrobial and edible. Birds, butterflies, and other wildlife use resin for survival. Humans use it for ink, lacquer, varnish, and other products.

Fireflies are really beetles. There are more than 2,000 species of fireflies, but not all are bioluminescent. Other bioluminescent organisms live in the ocean. Bioluminescent light is created by a chemical reaction during which oxygen combines with calcium, adenosine triphosphate (ATP), and luciferin with the help of the enzyme luciferase. Most fireflies that fly around at night are males looking for a mate. Once a female spots a male she likes, she responds with the same light pattern.

Female Ruby-throated hummingbirds don’t have ruby throats. Their throats are white, as are the tips of their tail feathers. Both the male and female young of this species look like females until their first molt. At that point, the male offspring have acquired their irridescent red gorget (throat).

Flowers have color adaptations for their specific pollinators. Some flowers use a “bull’s eye” pattern to stand out. Bees are attracted to blue, green, and ultraviolet light. Birds that are “ultra-violet-sensitive” will pollinate flowers from a larger portion of the color spectrum than those that are "violet-sensitive". Yellow flowers tend to be pollinated by flies. Butterflies pollinate a range of colors: white, pink, purple, red, yellow and orange.

<u>Flowers have scent adaptations for specific pollinators</u>. Bees are attracted to sweet or spicy scents. Bat-pollinated flowers usually produce strong fruity or musky scents while moth-pollinated flowers produce very sweet fragrances.

<u>Flowers have shape adaptations for specific pollinators</u>. Flowers that tilt downward, with long spurs containing nectar tend to be pollinated by hummingbirds. Flowers that have a "landing platform" tend to be pollinated by butterflies.

<u>Male bullfrogs croak to attract female frogs for breeding</u>. They also croak to warn other male frogs away from their territory. Female bullfrogs don't croak.

<u>Monarch butterflies are noxious due to the cardenolides (from *Asclepias syriaca*) they ingested as caterpillars</u>. The toxins remain in the insect's system, even after metamorphoses into a butterfly.

<u>Monarch and Viceroy butterflies are examples of Müllerian Mimicry, which is when unrelated noxious organisms resemble each other, with both benefiting</u>. Müllerian Mimicry is also found in some species of ants, bumble bees, and coral snakes.

<u>Mosses are small flowerless plants that typically form dense green clumps or mats,</u>

often in damp or shady locations. They are composed of simple leaves that are one cell thick, attached to a stem. They do not have a complex vascular system.

Moths and butterflies are in the same insect family (*Lepidoptera*), but there are a few differences. Butterflies generally rest with their wings closed, while moths rest with their wings open. Butterflies have thin antennas, while the antennae of moths are feathery. Butterflies generally gather food during the day while moths are active at night.

Old Snags are important structural components in wildlife communities. Dead and decaying wood supports a rich community of decomposers like bacteria and fungi, insects, and other invertebrates. These organisms and their consumers, along with the structural complexity of cavities, hollows, and broken tops make snags an important habitat for birds, bats, and small mammals, which, in turn, feed larger mammalian predators. The old snag shown in the meadow also provided necessary structure for The Trumpet Creeper.

Organisms produce more offspring than can survive. There are variations among organisms. Organisms compete for limited resources. Those organisms that are most fit for their environment will be able to survive long enough to reproduce and pass their genes on to their offspring. Species whose members don't survive long enough to pass on their genes to their offspring will go extinct.

Plants have different needs for sun, water, soil, and pH. You would not treat a black-eyed-Susan like a hydrangea. Black-eyed Susans need full sun, soil pH between 6-7, and can tolerate drought, if well established. Hydrangeas like partial sun, have variable flower color depending on soil acidity, and need consistent moisture.

Plants make noise. Researchers have found that tomato and tobacco plants made sounds when stressed by lack of water or when their stems were cut. Their sounds are in the ultrasonic range, which insects and some mammals can hear.

Plants might be able to hear. Flowers attract birds and insects with their nectar so that those animals will inadvertently transfer pollen and assist with plant reproduction. Some scientists have found that some flowers may sense the sound waves of nearby bee wings and increase the sugar concentration of their nectar in order to entice the pollinator to visit. *Echinacea* can change its scent.

Plants move. Time-lapse photography is a technique that can be used to show plants' life cycles. A photographer using the technique takes a bunch of photos of the same plant or group of plants over a period of time and then plays the images back as a movie. This way it is easier to see the process of plant growth and development.

Plants overproduce seeds because most of them won't survive. One cattail can have 25,000 seeds. A cottonwood tree can release more than 25 million seeds, each suspended by cottony fibers that allow it to ride on the wind. A dandelion seed head (puffball) can contain up to 172 seeds, but since the plant blooms repeatedly through summer, it can release 5,000 seeds a year. Most of those potential offspring are consumed, or used, by birds and other animals.

The Golden Garden spider is also called the Writing Spider. This spider's distinctive web is circular, with a diameter of up to two feet. There is a zig-zag pattern (writing) in the center. Golden spiders typically stay in the center of the web, waiting for prey to become ensnared. They consume the interior of their webs nightly and rebuild the next day with fresh silk.

There are 17,000 species of lichens. Lichens are composite organisms that arise from algae or cyanobacteria living among filaments of multiple fungi species in a mutualistic relationship.

There is no such thing as waste. Decomposers, such as earthworms, bacteria, termites, and even viceroy butterflies are necessary to break down organic materials. One organism's "waste" is another's food.

Thinking like a scientist means understanding the physical world using observable evidence, reasoning, and repeated testing. Developing new ideas about the way the world works comes from conducting tests and making observations. Investigating questions in a logical manner, as the meadow-ians were doing, is using the scientific method of inquiry.

Viceroys are mimics in all life stages. Their eggs resemble insect galls (growths from parasites) on host plants. Their caterpillars resemble bird droppings. Overwintering caterpillars roll themselves up in a leaf to hide from predators. As adult butterflies, Viceroys resemble monarchs.

Viceroys are noxious due to the salicylic acid they ingested (from Willow trees) as caterpillars. When Viceroys are stressed, they also release volatile phenolic glycosides, which deter predator attack.

Glossary of Sciency and Non-Sciency Words

Appendages- an additional part growing off of the main section; examples include arms, legs, antennae, etc.

Arachnid- an arthropod, such as a spider or scorpion

Coevolution- the influence of closely associated species on each other during their millions of years of development on earth

Deceitful- a misleading or false manner

Drone- a remote-controlled aircraft; make a low humming sound; can be very tiny

Forelimbs- front limbs of an animal

Frenemy- one who is friendly, despite dislike (being an enemy)

Incubate- keep eggs warm so they hatch

Insect- an animal with three body parts, three pairs of legs, and two pairs of wings

"Meadow-ians"- residents of the meadow

Mechanical- relating to machines or machinery

Mutualism- a type of symbiotic relationship where both organisms benefit

Perennial- plant that lives more than two years

Photosynthesis- a process used by plants and some other organisms to convert light energy into nutrients

Ponder- think about carefully

Proboscis- in invertebrates, a long mouth appendage; tongue

Respiration- a process where oxygen and nutrients are converted into energy (adenosine triphosphate (ATP))

Symbiosis- interaction between two organisms who live or associate closely

Unfurled- spread out; open; unrolled

Request for a Review

I sincerely hope you enjoyed using this educational tool as much as I enjoyed writing it. If you did, I would greatly appreciate a short review on Amazon or your favorite book or instructional materials website. Reviews are crucial for letting others know of the value of any instructional resource, and even just a line or two can make a huge difference.

Connect with The Author

G. Katz Chronicle has over three decades of teaching experience in primary, secondary, and higher education. Having taught all major content areas in primary and middle school, as well as education theory and biology at the secondary and college level, she most enjoys teaching the Life Sciences.

With a PhD in Curriculum and Instruction, she, too, is a perpetual student of traditional education and all areas of self-improvement.

G. Katz is an abbreviated form of Gertrude Katz. Gertrude maintains a Biology Teaching Blog where she provides her general instructional plans, reasoning, and thoughts for the NY State Living Environment Course. The blog consists of ten posts, which correspond to ten instructional units, and include unit presentations and videos. Gertrude invites you to check it out at: GertrudeKatzChronicle.com. Connect with her on other social media, as well!

Made in the USA
Las Vegas, NV
03 November 2026

57422981R00064